FANTASTIC PLANTS

KELP FORESTS

MARY GRIFFIN

PowerKiDS press
New York

Published in 2023 by The Rosen Publishing Group, Inc.
29 East 21st Street, New York, NY 10010

First Edition

Portions of this work were originally authored by Patricia Fletcher and published as *Kelp: The Underwater Forest!* All new material in this edition was authored by Mary Griffin.

Editor: Therese Shea
Book Design: Michael Flynn

Photo Credits: Cover, p. 9 divedog/Shutterstock.com; back cover, interior background (illustration) OlgaChernyak/Shutterstock.com; interior frame Liubou Yasiukovich/Shutterstock.com; p. 4 Zhane Luk/Shutterstock.com; p. 5 Ethan Daniels/Shutterstock.com; p. 7 SeaHamk77/Shutterstock.com; p. 10 NatalieJean/Shutterstock.com; p. 11 Aldona Griskeviciene/Shutterstock.com; p. 13 (fish in kelp) SHINART/Shutterstock.com; p. 13 (anemone) fatamorgana-999/Shutterstock.com; p. 15 Joe Belanger/Shutterstock.com; p. 17 Brandon B/Shutterstock.com; p. 18 Philip Garner/Shutterstock.com; p. 19 SeaHawk77/Shutterstock.com; p. 21 Aleksey Bakhtin/Shutterstock.com.

Library of Congress Cataloging-in-Publication Data

Names: Griffin, Mary, 1978- author.
Title: Kelp forests / Mary Griffin.
Description: New York : PowerKids Press, [2023] | Series: Fantastic plants | Includes index.
Identifiers: LCCN 2021051076 (print) | LCCN 2021051077 (ebook) | ISBN 9781538386552 (library binding) | ISBN 9781538386538 (paperback) | ISBN 9781538386545 (set) | ISBN 9781538386569 (ebook)
Subjects: LCSH: Giant kelp–Juvenile literature. | Kelp bed ecology–Juvenile literature.
Classification: LCC QK569.L53 G75 2023 (print) | LCC QK569.L53 (ebook) | DDC 577.3–dc23/eng/20211028
LC record available at https://lccn.loc.gov/2021051076
LC ebook record available at https://lccn.loc.gov/2021051077

Manufactured in the United States of America

Some of the images in this book illustrate individuals who are models. The depictions do not imply actual situations or events.

CPSIA Compliance Information: Batch #CSPK23. For Further Information contact Rosen Publishing, New York, New York at 1-800-237-9932.

CONTENTS

NOT QUITE PLANTS

Have you ever hiked through a forest? It's a great way to spend time in nature. You haven't hiked through at least one kind of forest—a kelp forest. You could swim through one, though! Kelp forests are important underwater **habitats**.

Kelp can grow really tall like trees can, but they are *not* plants. They're algae. Algae are living things that are mostly found in water. They're like plants but have some big differences. Let's learn more!

BE AN EXPERT!

"ALGAE" IS THE WORD FOR MORE THAN ONE "ALGA."

You might call kelp "seaweed." Seaweed is the name for algae that grow near shores.

WHERE'S THE KELP?

Kelp forests are found in cool parts of Earth's oceans. Well-known examples of kelp forests exist off the coasts of South Africa, Australia, New Zealand, and the southern coast of South America. Kelp forests live near the western shores of North America too.

For kelp to grow as tall as trees, it needs deep water. However, it's not often found in water deeper than 130 feet (40 m). Kelp likes to grow close to other kelp, in large groups.

BE AN EXPERT!

KELP FORESTS ARE FOUND WHERE WATER HAS A LOT OF **NUTRIENTS**.

THE LARGEST KELP FORESTS ARE FOUND IN WATERS BELOW 68°F (20°C).

LET'S LOOK AT KELP

Unlike most plants, kelp doesn't have roots. At its base, it has a holdfast, which sticks to a rock or **reef** and keeps the kelp in place. Kelp also has a stipe that holds it up, which is like a plant's stem. Blades that look like leaves grow off the stipe.

Though kelp parts are different from plant parts, kelp makes food the same way. Its blades use sunlight to produce food through photosynthesis. Kelp absorbs, or takes in, nutrients from the water too.

BE AN EXPERT!

PHOTOSYNTHESIS IS THE NAME FOR HOW A PLANT OR PLANTLIKE **ORGANISM** TURNS SUNLIGHT, WATER, AND THE GAS CALLED CARBON DIOXIDE INTO FOOD FOR ITSELF.

ALGAE BLADES MIX UP THE WATER, BRINGING NUTRIENTS CLOSER.

Kelp has another interesting body part that plants don't: floats! On kelp, tiny **bladders** filled with gas act like floats. They make kelp's blades float up, closer to the water's surface. They help the algae absorb as much sunlight as possible. They also keep the kelp upright in the water. The air-filled pockets mostly contain the gases oxygen, nitrogen, and carbon dioxide.

Different kinds of kelp have floats in different places. Some kelp have just one large float. Others have many.

BE AN EXPERT!

THE NAME FOR THE AIR-FILLED BLADDER, OR FLOAT, ON KELP IS A PNEUMATOCYST.

KELP UP CLOSE

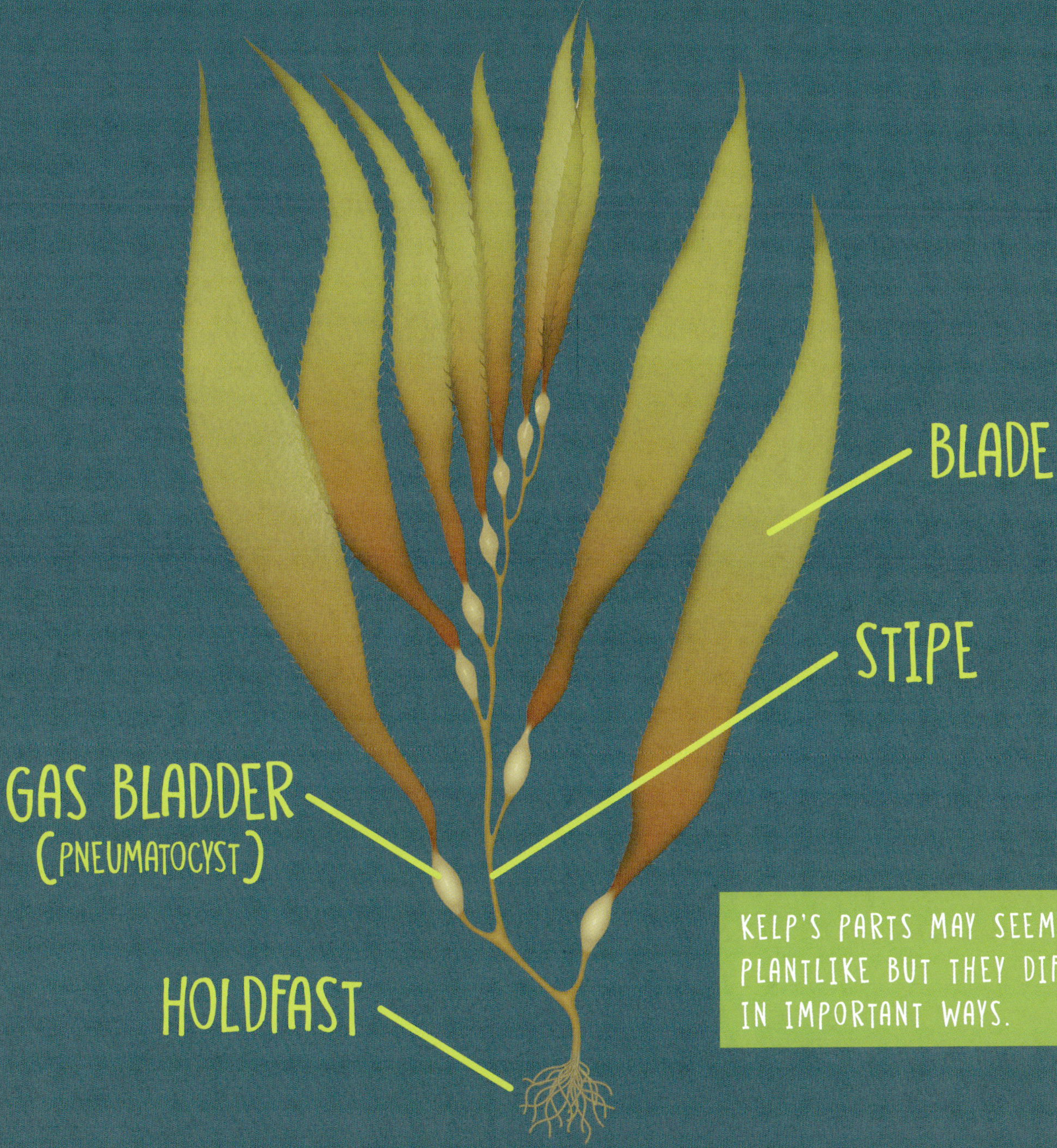

GAS BLADDER
(PNEUMATOCYST)

BLADE

STIPE

HOLDFAST

KELP'S PARTS MAY SEEM PLANTLIKE BUT THEY DIFFER IN IMPORTANT WAYS.

TAKE A TOUR

The top level of a kelp forest is the canopy. The canopy is where the tallest blades of kelp meet. Living in the canopy—and sometimes eating it—are crabs, snails, young fish, and other smaller ocean animals.

The name for the area under the canopy is the understory. Predator fish like bass and rockfish live there. They sometimes swim up to eat the smaller fish. The bottom of the kelp forest is called the forest floor. It, too, is a home for many organisms, including sponges and anemones.

BE AN EXPERT!

THE TALLEST KELP GROWS TO THE SURFACE AND THEN GROWS **HORIZONTALLY** ON TOP OF THE WATER.

THIS BASS LIVES IN A KELP FOREST OFF THE COAST OF CALIFORNIA.

PLACE OF PROTECTION

Kelp forests are a perfect place for sea creatures to find a meal. However, they also get **protection** hiding among the blades of algae. Kelp forests give fish, seals, otters, and even seabirds cover from storms and predators. They also act as nurseries, or places where young creatures can safely grow.

Kelp forests are helpful to people too. They protect shorelines from **damaging** waves. People eat many kinds of fish and shellfish that live in kelp forests, such as abalone, snapper, and lobster.

A SEAL SWIMS THROUGH A KELP FOREST LOOKING FOR A MEAL.

IN THE FOOD WEB

Kelp is an important part of ocean food webs. Some animals **rely** on the algae itself as a food source. Sea urchins eat kelp holdfasts, for example. Groups of urchins can really do damage if left unchecked. However, sea otters eat sea urchins. And in turn, larger sea creatures, like sharks, eat otters.

In the past, people killed too many sea otters for their fur. Without them, the sea urchin population grew and ate too much kelp. It took years for kelp forests to return.

BE AN EXPERT!

A FOOD CHAIN IS THE ORDER IN WHICH LIVING THINGS NEED EACH OTHER FOR FOOD. MANY FOOD CHAINS CONNECT TO FORM A FOOD WEB.

THE PURPLE SEA URCHIN POPULATION HAS GROWN NEAR CALIFORNIA, AND THE FORESTS OF KELP NEAR THE COASTLINE ARE DISAPPEARING.

FAST FORESTS

Kelp forests are mostly made up of giant kelp and bull kelp. Giant kelp is the largest **marine** algae and one of the fastest-growing living things on Earth. It can grow 2 feet (61 cm) in a day in the right conditions. It can grow to be 215 feet (66 m) tall in all!

Bull kelp is a fast grower too—up to 10 inches (25 cm) in a day. Its blades grow at the top of the algae, creating a thick canopy at the top of the water.

BULL KELP

BE AN EXPERT!

GIANT KELP IS FOUND NEAR COASTS FROM CALIFORNIA TO MEXICO AND NEAR SOUTH AMERICA, NEW ZEALAND, AND AUSTRALIA.

KELP GROWS BEST IN TURBULENT, OR CHOPPY, WATER. NUTRIENTS ARE ALWAYS PASSING BY FOR IT TO ABSORB.

WE NEED KELP!

Anything that changes water affects kelp forests. Polluted water can keep kelp from getting enough sunlight for photosynthesis. Pollution also affects nutrients in the water, which can harm kelp's growth.

Global climate change is the name for the change in Earth's weather patterns over time. Climate change is linked to rising temperatures and dangerous weather. Kelp forests store carbon dioxide, a gas that **contributes** to climate change. So, healthy kelp forests have an important role in protecting our planet!

BE AN EXPERT!

OVERFISHING CAN HARM KELP FORESTS. TOO MANY OR TOO FEW OF ANY ANIMAL OR PLANT IN A HABITAT CAN UPSET A FOOD WEB'S BALANCE.

PEOPLE USE KELP FOR ALL SORTS OF THINGS: FOOD, TOOTHPASTE, SHAMPOO, AND MORE.

GLOSSARY

bladder: A gas-filled part of certain algae that serves to help hold up the algae by floating.

contribute: To be one of the causes of something.

damaging: Causing harm.

habitat: The natural place where an organism lives.

horizontally: In a way that is level with the line that seems to form where the earth meets the sky.

marine: Having to do with the sea.

nutrient: Something a living thing needs to grow and stay alive.

reef: A chain of rocks or coral, or a ridge of sand, at or near the water's surface.

organism: A living thing.

protection: Something that keeps something else from being harmed.

rely: To depend on.

FOR MORE INFORMATION

BOOKS

Bate, Mathew. *With a Little Kelp From Our Friends: The Secret Life of Seaweed.* Port Melbourne, Australia: Thames & Hudson, 2021.

Blanchette, Carol, and Jenifer Dugan. *The Golden Forest: Exploring a Coastal California Ecosystem.* Lanham, MD: Muddy Boots, 2017.

Green, Jen, et al. *Oceans: Explore the Ocean's Layers from the Surface to the Seafloor.* San Diego, CA: Silver Dolphin Books, 2018.

WEBSITES

Giant Kelp
www.montereybayaquarium.org/animals/animals-a-to-z/giant-kelp
Find out more cool facts about these huge algae!

Kelp Forest
ny.pbslearningmedia.org/resource/kqed07.sci.life.eco.kelp/kelp-forest/
Swim through a kelp forest with this video.

INDEX